Other books in this series:

Spinning
Dyeing and Printing
Yarn Preparation
Fabric Manufacture

T0272873

MEDICAL AND HYGIENE TEXTILE PRODUCTION

Practical Action Publishing Ltd
27a Albert Street, Rugby, CV21 2SG, Warwickshire, UK
www.practicalactionpublishing.org

© Intermediate Technology Publications Ltd, 1994

First published 1994\Digitised 2008

ISBN 10: 1 85339 211 1
ISBN 13: 9781853392115
ISBN Library Ebook: 9781780444147
Book DOI: http://dx.doi.org/10.3362/9781780444147

All rights reserved. No part of this publication may be reprinted or reproduced or
utilized in any form or by any electronic, mechanical, or other means, now known
or hereafter invented, including photocopying and recording, or in any information
storage or retrieval system, without the written permission of the publishers.

A catalogue record for this book is available from the British Library.

The authors, contributors and/or editors have asserted their rights under the
Copyright Designs and Patents Act 1988 to be identified as authors of their respective
contributions.

Since 1974, Practical Action Publishing has published and disseminated books and
information in support of international development work throughout the world.
Practical Action Publishing is a trading name of Practical Action Publishing Ltd
(Company Reg. No. 1159018), the wholly owned publishing company of Practical
Action. Practical Action Publishing trades only in support of its parent charity
objectives and any profits are covenanted back to Practical Action (Charity Reg. No.
247257, Group VAT Registration No. 880 9924 76).

Cover picture: Allison Mathews
Typeset and illustrated by Ethan Danielson, Barnstaple, UK

MEDICAL AND HYGIENE TEXTILE PRODUCTION

A handbook

Allison Mathews and
Martin Hardingham

Intermediate Technology Publications 1994

CONTENTS

ACKNOWLEDGEMENTS

It would be impossible to mention all of the people who have helped in the research and production of this handbook. We would like to offer particular thanks to John Foulds, Dr John Gerrard and Dr Bill Cooke for their technical advice. We would also like to thank Mr B. Dayawansa Perera and Mr S. Wijetilleka in Sri Lanka, and the weavers of Nalta, Bangladesh who provided essential information; and Ethan Danielson for his artwork. Finally, we would like to thank the Thrasher Research Fund for their generous financial support which made the whole project possible.

Allison Mathews
Martin Hardingham

FOREWORD

This handbook is one of a series dealing with small-scale textile production, from raw materials to finished products. Each handbook sets out to give some of the options available to existing or potential producers, where their aims could be to create employment or sustain existing textile production, the ultimate goal being to generate income for the rural poor in developing countries. Needless to say, this slim volume does not pretend to be comprehensive. It is intended as an introduction to the topic which may stimulate further enquiry. Although each handbook is complete in itself and provides useful reference material on each specific area of production, the series, taken as a whole, reveals the breadth of technology required to equip a small-scale textile industry. While being primarily technical, the series also covers some of the socio-economic, managerial, and marketing issues relevant to textile production in the rural areas of developing countries.

Production of this series of books has been sponsored by Intermediate Technology as part of its effort to help co-ordinate the most appropriate solutions to particular development problems. The series forms part of the cycle of identifying the need, recognizing the problems, and developing strategies to alleviate the crisis of un- and under-employment in the South.

Intermediate Technology also offers consultancy and technical enquiry services. If you require further information, we will be pleased to help.

Martin Hardingham
Intermediate Technology, Rugby, UK

PREFACE

The Medical and Hygiene Textiles Project, which gave rise to the production of this handbook, began in 1989. Intermediate Technology (IT) had received numerous enquiries from developing countries around the world for information on the small-scale production of bandages, gauze, cotton wool, sanitary towels and nappies. In response to these enquiries, IT conducted a desk study of existing small-scale technologies, which culminated in an international forum held at the London School of Hygiene and Tropical Medicine. Following the forum, IT concluded that more research on the use, availability and small-scale production of medical and hygiene textiles was needed. This handbook has been produced to make the results of the research more widely available.

Intermediate Technology conducted technical and market research in Zimbabwe, Bangladesh and Sri Lanka, and tapped technical and medical expertise available in Europe. This research revealed that the production of medical and hygiene textiles is often surrounded by a complex of infrastructural, cultural and health issues. For example, the market research in Zimbabwe revealed that although remote rural clinics do not have sufficient supplies of bandages, the production and distribution are centrally controlled, so that local production would not be able to satisfy needs anyway. In spite of a perceived need for these products, therefore, small-scale production may not always be an appropriate solution.

This handbook outlines medical and hygiene textile production so that rural communities and development organizations or individual field workers involved with them can learn a little about the principles and processes involved, and find out where further help or knowledge is available. There is also basic information on market research, product specification, finishing and small-scale textile manufacture.

Allison Mathews
Intermediate Technology, Rugby, UK

1. INTRODUCTION TO MEDICAL AND HYGIENE TEXTILES

This handbook describes the processes which can be undertaken by hand or by the use of small, mechanically driven equipment. All production stages, from raw material processing to the packaging and labelling of the finished products, are covered.

The term medical and hygiene textiles refers to all textiles which are used for first aid, clinical or hygienic purposes. In this handbook, medical textiles include gauze, bandages, and cotton wool. Hygiene textile products consist of sanitary towels (napkins) and nappies (diapers).

In many countries of the South, locally available materials are often used for medical purposes, such as absorbent plant fibres to cover and protect wounds. Cotton rags are used for sanitary protection and traditional swaddling clothes for babies' hygiene. Although these alternatives are not covered in this handbook, it is accepted that the knowledge of traditional, indigenous materials and their use is important and valuable.

Historically, cotton was the prime raw material for making conventional medical and hygiene textiles. While cotton will always be used as a raw material throughout the world, its relative overall importance is declining, particularly in countries of the North. Conventional woven cotton dressings, for instance, are being replaced by non-woven fabrics. Fibres made from wood pulp have virtually replaced cotton in the production of sanitary towels. Cotton medical and hygiene textiles continue to be manufactured and used in developing countries, however.

Medical and hygiene textiles are supplied in developing countries from a variety of sources. They can be manufactured locally using both large- or small-scale technologies. Sometimes they are wholly or partially imported. In many countries production and distribution are controlled centrally.

TYPES OF MEDICAL TEXTILES

There are five basic classifications of medical textiles. Products mentioned in *italics* are not covered in this handbook.

Absorbents
(Absorbent cotton, gauze swabs). These absorb moisture without contaminating the contact area with unwanted substances or bacteria. Cotton wool is often used to clean a wound or for applying antiseptic, and gauze swabs are generally used during surgery to absorb blood or discharge. As with all medical textiles, it is essential that these are sterile.

Wound dressings
(Gauze, *adhesive bandages*). A dressing is a protective covering which is placed on a wound to promote healing, to prevent contamination and infection, to absorb discharge, and to avoid further injury. A dressing should be absorbent and porous, to allow the skin to breathe, and should not stick to the wound.

Dressing retention
(Bandages, *adhesive tapes*). These serve to hold dressings in place.

Support bandages
(Crêpe bandage, calico, knitted tube bandage). Support bandages hold splints in position, provide support to the body, limbs, or joints evenly, and they restrict movement. They should be lightweight, and cause no side-effects such as skin rashes.

Paddings
(Fibre pads, *adhesive felt, orthopaedic bandages*). Paddings provide relief of pressure and friction to a part of the body, for example, under a plaster cast.

TYPES OF HYGIENE TEXTILES

Sanitary protection
(Sanitary napkins, absorbent cotton, *rags*). These products are used to absorb menstrual blood, and are also used following childbirth. Sanitary towels produced from wood pulp fibres and non-woven materials require large-scale, industrial machinery, and are not covered in this handbook.

Nappies (diapers)
These are used to absorb waste from babies. Non-disposable nappies are generally made from terry or muslin fabric. *Disposable nappies* are manufactured from wood pulp fibres.

NON-TECHNICAL CONSIDERATIONS

There are a number of non-technical issues which should be taken into account if medical or hygiene textile production is being considered. The key elements are needs and markets.

It is important to identify the need for a product, and to understand why the need exists. There may be a number of factors which contribute to need, and production may not always be the most appropriate way of addressing it. The need for medical textiles may not necessarily imply that there is a market. For example, the need for improved sanitary protection might be identified among a particular group of women, but financial constraints might limit their ability to purchase sanitary towels.

If commercial production is being considered, then there should be a full understanding of the markets. Commercial production has the benefit of offering opportunities for employment and income generation, as well as fulfilling needs. It essential that a thorough market feasibility study be conducted before any type of production is initiated. Market feasibility and other aspects of production planning are addressed more fully in Chapter 5.

In general, medical textiles differ from hygiene textiles in terms of their use. This difference is likely to influence decisions on how the needs should be addressed, potential markets for the products, and whether small-scale production would be a viable option. Dressings and bandages are most often used in hospitals and clinics, with comparatively few being used in individual households. Sanitary towels and nappies are more commonly used on an individual household level, the exception being maternity hospitals.

Generally, women in developing countries use the materials most available and affordable for sanitary protection. This is often a piece of absorbent material, held in place with a string or cloth belt, which is washed and reused. Absorbent cotton is also commonly used. Disposable sanitary towels may be used by urban, middle-class women.

The types of baby nappies used throughout the world vary greatly. Disposable varieties made from wood pulp are favoured in the North and among urban middle-class households in the South. Nappies may not be used at all in some countries.

Factors affecting the use of medical textiles

HEALTH The use of poor quality bandages or dressings, or in some cases, of indigenous alternatives, may cause infections. Serious occurrences of infection may constitute a need for a better quality product.

ECONOMIC Economic and infrastructural factors may also affect the use and availability of medical textiles. Insufficient supplies to hospitals and clinics may be caused by a shortage of funds, distribution or transport problems. A more immediate need, therefore, may be improved infrastructure, and/or delivery systems.

Factors affecting the use of hygiene textiles

CULTURAL In most traditional societies, menstruation is surrounded in rituals and is rarely discussed in public. Myths and taboos often limit women's mobility and daily activities during menstruation. They also contribute to a feeling of shame and embarrassment. Anxiety about 'leaking' may affect a woman's attendance at work or school during menstruation.

The issues surrounding menstruation and sanitary protection are numerous and complex. A thorough investigation of existing practices should be undertaken before decisions about production are made. Activities such as health education and awareness-raising may be more appropriate ways of addressing the problems.

An investigation into sanitary protection practices should be conducted with sensitivity. Methods for gathering information on cultural aspects of sanitary protection practices are covered more fully in Chapter 5, and in Appendix 1.

HEALTH Use and care of traditional sanitary protection materials such as rags sometimes cause health problems. Cultural taboos may require women to dry their rags inside, away from the sanitizing effects of the sun. Sometimes there is a shortage of clean water or soap. Rags washed and dried in such conditions have been known to contribute to rashes and reproductive tract infections.

ECONOMIC In many developing countries, disposable sanitary towels are only affordable by urban, middle-class women. Poor, rural women often cannot afford sanitary towels, and may not consider them important enough to spend money on. Disposable, commercially produced sanitary towels may not be available in rural areas.

SMALL-SCALE PRODUCTION

This handbook gives technical information and specifications for the production of a range of textile items on a small scale. Only a small number of basic items are required to perform the various functions intended.

In the North a wider variety of medical and hygiene products is available because of mass-production techniques. Many of these products are manufactured from materials such as fibres from wood pulp and alginate. The production of these materials requires sophisticated technologies and involves high capital expense. For these reasons they are not appropriate for small-scale production.

Medical and hygiene textiles need to be relatively inexpensive to produce because they are only used once. For this reason, large-scale, capital intensive methods are most appropriate for their production. Small-scale production is economically viable in many countries where labour costs are low, and there are small markets. Therefore it is important that production costs are assessed carefully before decisions about production are finalized.

Most textile items which are used in hospitals and clinics in developing countries are made of cotton. This handbook therefore focuses primarily on cotton production, but does not preclude in certain circumstances the use of local indigenous plant fibres.

The initial stages of production of medical and hygiene textiles can be categorized as follows.

1.   Selection of raw materials

2.   Fibre processing

3.   Spinning

4.   Fabric production

5.   Fabric finishing

6.   Fabric conversion

7.   Packaging and labelling.

Chapter 2 of this handbook describes the initial production processes, from raw material processing to fabric production. Chapter 3 provides information on how textiles are made specifically for medical purposes. Chapter 4 tackles the important subject of quality control, and Chapter 5 introduces the equally essential considerations of marketing and commercial production. Chapters 6 and 7 provide information on equipment suppliers and further resources.

2. PRINCIPLES OF TEXTILE MANUFACTURE

As can be seen in Chapter 1, most textile products follow similar initial stages of manufacture. The difference between medical and hygiene textiles and other textile products is in the way they are used.

SELECTION OF RAW MATERIALS

Consideration must first be given to the various types of fibre from which textiles are made. Medical and hygiene textiles have a number of important requirements which determine which fibre is suitable for their manufacture. The criteria which determine the selection of suitable raw materials can be summarized as follows.

Use
The intended use of a medical textile product is the most important consideration for fibre selection. For example, the fibres should be absorbent. However, many fibres, such as cotton, only become very absorbent after some processing.

Availability and cost
The fibre should be readily available and relatively cheap to produce, cultivate, harvest, transport and process.

Staple length
If fabric is to be produced, the fibre should be medium to long for spinning into a yarn suitable for weaving or knitting. If an absorbent wadding is the end product, shorter staple fibre can be used.

Comfort, purity
The fibre should be soft and smooth so as not to cause irritation. It should also be relatively free from additives, contaminants and residues, which may also cause irritation.

Bio-degradability, burnability
The fibre should preferably be bio-degradable or burnable.

Cellulosic fibres
Staple cellulosic fibres are the most acceptable and most suitable raw materials for the production of medial and hygiene textiles, because they have most of these characteristics. Staple cellulosic fibres are mostly derived from the seeds, stems, or leaves of certain plants, for example:

Seed fibres: cotton, kapok

Bast fibres: flax, ramie, hemp or nettle, jute

Leaf fibre: banana leaf, sisal, manila

Other cellulosic fibres, such as viscose rayon, can be manufactured from wood pulp.

Cotton is the most widely used raw material for the production of medical and hygiene textiles in most developing countries. The advantages of using cotton are that it is still economical to process, and it is often grown locally.

Cotton

Cotton is a soft fibre which grows from the seed pod of the cotton flower. Each cotton fibre is a tube of almost pure cellulose. Raw cotton fibres possess a natural protective wax coating which repels water. During the course of fibre processing, fabric production and finishing, these natural waxes and oils are removed and the cotton then has excellent moisture absorption capacity.

Cotton is grown in more than 90 countries of the temperate and tropical regions of the world. It is a warm climate plant and the cultivated species are not tolerant of freezing temperatures. Cotton needs a reasonable amount of water to grow. In many countries, this requirement may be adequately met by rainfall, but cotton is also grown in arid and semi-desert conditions, such as in Sudan, where vast cotton plantations are irrigated. Methods of cultivation vary considerably from country to country. Although much of the cotton grown in the world is planted, maintained and harvested by machinery, cotton is still cultivated, tended and picked by hand in some countries on small, family-managed plots.

Of the many varieties of cotton grown throughout the world, there are three basic types of fibre. Each type is used mostly to make specific products. The quality of cotton fibre is determined by its staple length, fineness, uniformity, maturity, colour, and cleanliness. Cotton is classified into three main groups:

Long: high-lustre fibre with $1\frac{1}{4}$ in.–$2\frac{1}{2}$ in. (30–65mm) staple length: Sea Island (West Indies, Central America, and Mexico), Egyptian, Sudanese, Peruvian, American Pima, East African.

Medium: $\frac{3}{4}$ in.–$1\frac{1}{2}$ in. (20–30mm) staple length: American Upland (the bulk of US production)

Short: coarse, dull $\frac{1}{2}$ in.–$\frac{3}{4}$ in. (10–20mm) staple length: Indian, Chinese.

Medium and short staple cotton are suitable varieties for the production of many medical textile products such as absorbent cotton, sanitary towels and absorbent gauze.

Other vegetable fibres

Other vegetable fibres which have absorbent properties, such as flax, allo nettle, ramie, and a range of other bast and leaf fibres, could be used for medical and hygiene textile production. The economic and technical aspects of processing alternative fibres require further research (see Chapter 7).

FIBRE PROCESSING

An initial processing stage is required to separate fibres from the rest of the plant. In the case of cotton this process is known as *ginning*. Bast fibres are extracted from the stem of a variety of plants by *retting* and *scutching*. Retting is soaking or fermenting the plant in water to loosen the fibres. Scutching is the process of separating the fibres from the woody core. Leaf fibres are obtained by scraping, retting and scutching.

The fibres which remain after the initial processing are called staple fibres. As a result of the initial processing the staple fibres are matted together in a haphazard way. In order to use the fibres in further textile production they must be put into an orderly arrangement. Most fine staple vegetable fibres can be further processed in similar ways to cotton.

Although small-scale fibre processing is possible, it is generally more economically viable on a large scale. The following processes are used to convert a mass of fibres into yarn.

❑   Opening, cleaning, and mixing fibres.

❑   Carding the opened mass of fibres so that they are more or less parallel, forming a rope of fibres called a *sliver*.

❑   Spinning the fibres by drafting and extending them over each other until a desired thickness is achieved and inserting twist to form a yarn. Yarns are then wound on to a bobbin or another type of package.

COTTON PROCESSING

Opening and cleaning
Cotton is either hand- or machine-picked in the field and then ginned before it is compressed into bales. The bales are then transported to a store or textile mill for processing. Illustration 1 shows the pre-spinning process.

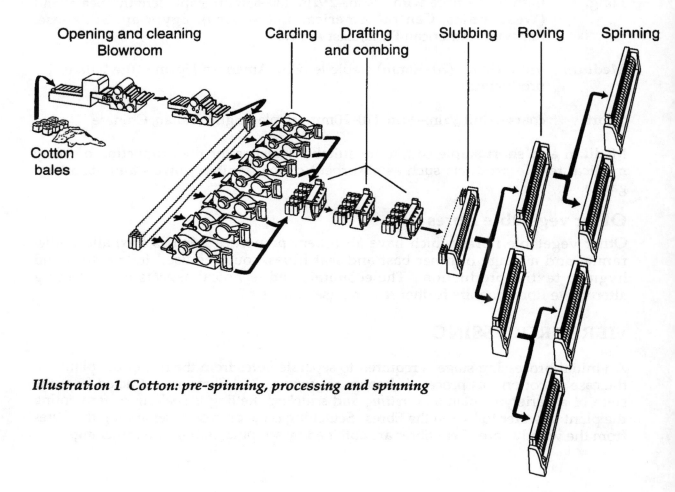

Illustration 1 Cotton: pre-spinning, processing and spinning

At this stage, if absorbent cotton (cotton wool) is to be produced, the cotton receives a wet treatment to remove the natural wax and to clean and bleach the fibre. The clean, bleached and dried fibres are then opened and carded. The production of absorbent cotton is covered in more detail in Chapter 3.

Cotton intended for other textile products is opened and cleaned. This initial stage is carried out in the blowroom where the cotton is mechanically opened into a loose, fluffy mass. Trash and other impurities gathered during picking are removed and the fibres are arranged roughly into a thick blanket which is then rolled into a *lap* (see Glossary).

Carding

The lap is passed through a machine known as a *card* which disentangles the fibres, removes remaining impurities and begins to arrange the fibres in one direction. The carding process produces a smooth, even web of fibres which is then gathered together and loosely twisted into a rope of fibres of about 25mm in diameter — the sliver.

Combing

This is only used to process longer, finer fibres, which are arranged even more uniformly and parallel, to produce higher quality yarn.

Drawing

These final processes prepare the sliver for spinning, by combining and drawing out the fibres through a *draw frame*, which produces a sliver with more evenly arranged fibres. The sliver is then drawn even further through a *speed frame* into a finer strand, and is slightly twisted. This strand of cotton fibres, known as a *roving*, is now ready for spinning.

Spinning

There are three essential actions of spinning.

1.   Drawing fibres from a sliver or roving, reducing the fibres to the final thickness which determines the *count* of the finished yarn.

2.   Inserting twist to hold the fibres in place in order to form a yarn, and determine the softness or hardness of the yarn (see Illustration 2).

3.   Winding the yarn on to a bobbin or another type of package.

Folding, doubling and twisting

In order to make an even stronger yarn a single yarn can be combined with another, thus forming a two-fold yarn. This operation is called either folding, doubling or twisting.

Yarns are folded together with the opposite twist to that used in the spinning process. It is normal to use Z-twist in spinning. S-twist is usually used in the twisting operation (see Illustration 3).

Sizing

Size is often applied to warp yarns made from single yarns to reduce friction and breakage during weaving. There is a range of commercially available sizing agents, but size is often made from natural starches derived from sago, maize, tapioca, farina or rice.

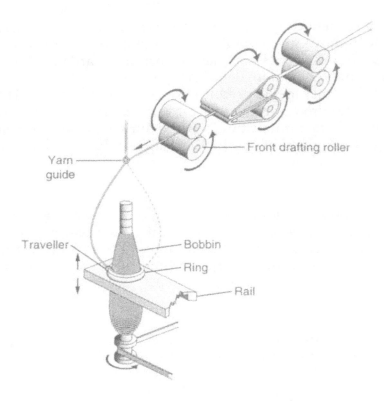

Yarn guide

Front drafting roller

Traveller

Bobbin

Ring

Rail

Illustration 2 Ring spinning

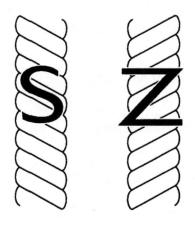

Illustration 3 S-twist and Z-twist

Before a yarn can be woven or knitted into a fabric it goes through a number of preparatory stages. These processes are outlined in Chapter 3 and described in detail in *Yarn Preparation* (see Chapter 7).

YARN SPECIFICATIONS

Different medical and hygiene textile products will require different specifications of yarn. The type of yarn to be used depends on the fabric production process, such as weaving or knitting.

Yarn counts

The *count* of yarn is a number which expresses its thickness. The count is defined as the weight per unit length (direct yarn count systems), or the length per unit weight (indirect yarn count systems).

Many yarn count systems have come into use over the years, all using different units of length and weight. In this handbook all counts are given either in the Tex (direct fixed-length) or the standard English Cotton Count system (indirect fixed-weight).

In the Tex system, the fixed unit of length is 1000 metres and the weight is in grams (g). For example if 1000 metres of yarn weighs 30g, the yarn is 30 Tex.

In the English Cotton Count system the fixed unit of weight is one pound. The count is the number of 840 yard lengths in one pound. For example, the count of one pound of yarn 16 800 yards (i.e. 20 X 840) long would be 20s cc (cotton count).

Twist

The amount of twist in a yarn plays a very important part in determining its character, in particular its hardness or softness and strength, are all determined to a large extent by the amount of twist.

The amount of twist is decided by the end-use of the yarn. Warp yarns, for example, which need to be strong and elastic to withstand the strains of weaving, are given more twist than weft yarns, which are not normally subjected to much tension and need to be softly twisted and bulky to give good cover. For special qualities in the yarn, the twist used may be more than that necessary to provide maximum strength, for example in yarns for crêpe fabrics.

FABRIC PRODUCTION

There are two basic methods of fabric production used in medical textile manufacture: weaving (interlacing) and knitting (looping).

Weaving

Woven fabrics have two sets of yarns interlacing at right-angles: the warp yarns, or ends, which lie along the length of the fabric, and the weft yarns, or picks, which are inserted across the width of the fabric one after the other as weaving proceeds. The variety of arrangements of yarns in a woven structure can be very wide. Plain weave is the simplest method of interlacing, and is the most commonly used for medical and hygiene textile production (see Illustration 4).

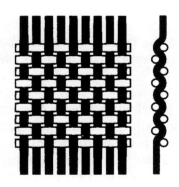

Illustration 4   Plain weave

Weaving can be carried out using simple or sophisticated looms. One advantage of weaving is that even complicated looms can be made from basic materials such as wood and can be constructed locally.

Illustration 5 shows the working parts of a simple two-shaft loom. The warp yarns are wound side by side on the warp beam and then taken over the back beam. Next, the lease rods are inserted through the warp yarns to maintain their order and to ensure that they do not become entangled. The warp ends are entered individually through the healds, then through a reed, over the front beam and finally secured to the fabric roller at the front of the loom. The warp yarns are continually held under tension between the warp beam at the back and the cloth beam at the front.

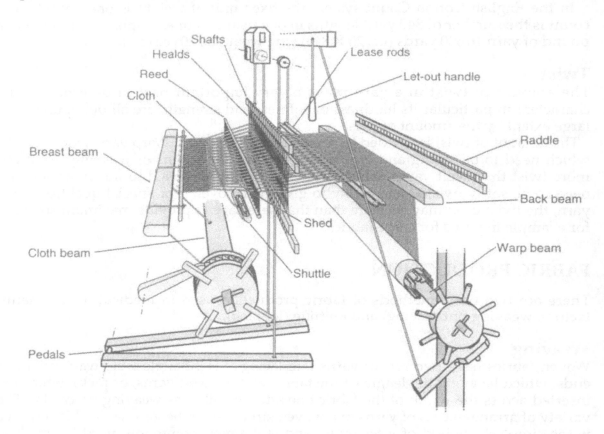

Illustration 5 Working parts of a simple two-shaft loom

Narrow width looms such as the inkle loom could be used for weaving bandages, giving a woven edge, or selvedge, on each side of the bandage.

Knitting

Knitted fabrics are made from interlocking loops of yarn. There are two distinct types of knitted fabric: weft-knit, so-called because the yarns lie across the fabric, and warp-knit, in which the yarn path is basically down the length of the fabric.

Each row of stitches is called a *course*, and each column is called a *wale*. In machine knitting, each single wale of stitches will be made on a separate needle. Knitted fabrics made with knitting needles, by hand or on domestic knitting machines, are usually weft-knit. Weft-knitting differs from weaving and warp-knitting in that the weft-knit fabrics can be made from a single yarn package. Warp-knit fabrics are almost invariably made on complex machines for a limited number of end-uses. Knitting is covered more fully in *Fabric Manufacture* (see Chapter 7).

Machine weft knitting

The knitting machine consists of a needle bed containing a large number of needles, one for each wale (column of stitches) of the fabric. One course is knitted by moving a carriage delivering yarn across the needle bed in one operation, so the actual speed of knitting is considerably faster than hand-knitting.

A hand-knitting machine is one in which the movement of the carriage is brought about by hand; the power machine has a motor to run it, but it works on the same principles. The carriage which delivers the yarn to the needles also contains a cam system which controls the operations of the needles. There are several types of needle, but in the vast majority of weft knitting machines the latch needle is used (see Illustration 6).

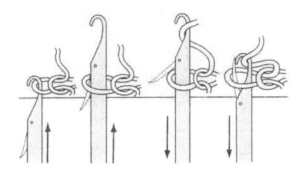

Illustration 6 Latch needles used in weft knitting machines

The needle movement is vertical, controlled by the movement of the cam as it passes across. The stitch is formed as follows. The needle rises to allow yarn to enter the needle head; it then drops, pulling the yarn into a loop and at the same time the previous loop slips over the needle latch and clears the needle. The stitch cam on the carriage can be adjusted to set the distance the needle moves down and this therefore controls the length of the stitch.

Knitting can be suitable for the production of some medical and hygiene textile items, such as bandages and sanitary towels.

Specifications of weft knitting machines

Knitted fabric can be made by hand by using two needles, but this method is highly unproductive compared with using a knitting machine. On the other hand, knitting machines, although fast and fairly simple to use, are complicated pieces of equipment. The production of latch needles for these machines is only viable on a commercial basis, so for small-scale production even the simplest knitting machines have to be bought.

Hand V-bed machine

This hand-knitting machine with two needle beds of latch needles is shown in Illustration 7. A carriage carrying the yarn is moved by hand across the bed of the machine. Each movement creates one course of knitting.

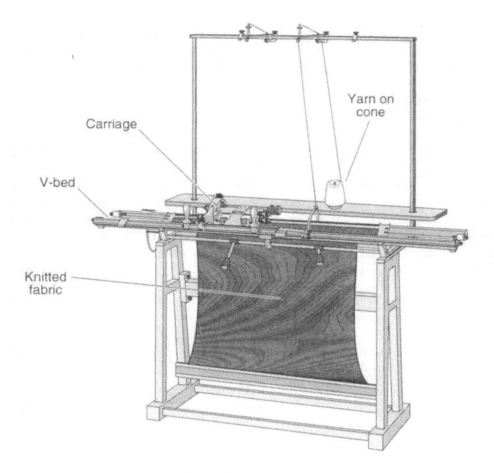

Illustration 7 Hand V-bed weft knitting machine

Circular hand-knitting machine

Illustration 8 shows a small hand-knitting machine for tube knitting. It is constructed from metal with latch needles. It will only produce a single-width tube. Its output is approximately 30 metres of plain knitted tube per hour.

WET PROCESSING

Cotton cloth, straight from the loom or knitting machine, cannot be used without further treatment which is generally referred to as wet processing. These processes may include removing any size (de-sizing); scouring, to clean the cloth and remove impurities such as natural waxes and seeds; and bleaching, which destroys coloured substances and adds an even whiteness to the cloth. They are described in more detail in Appendix 3.

  Following wet processing the dry fabric is rolled or folded or cut to size. Packaging in paper or plastic to ease transport and to protect the fabric from dirt, and labelling the package for marketing are the final stages.

TESTING AND QUALITY CONTROL

Quality testing and quality control should be an integral part of production. This is particularly important when a fabric is being manufactured to standard specification.

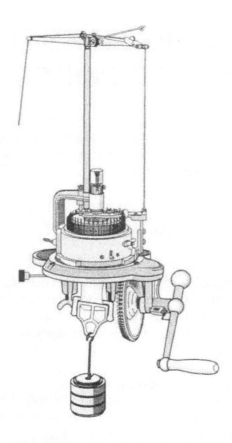

Illustration 8 Hand-operated circular knitting machine

'Quality control' means the procedures, actions and supervision needed to make a textile which meets a desired specification.

Testing is carried out to discover whether a specified quality standard has been achieved. This procedure also provides information which can be used to make changes in production to improve quality and prevent any variations from a desired specification.

In the production of medical textiles, testing is especially important, as a poor quality product may affect the health of the user. Most countries follow national or international standards for all medical textiles used (see Chapter 7).

Simple methods of testing medical and hygiene textiles are covered in more detail in Chapter 4.

3. MAKING TEXTILES FOR MEDICAL PURPOSES

The production of medical and hygiene textiles follows the same initial manufacturing process as other textiles. There are, however, two production considerations which differ from other processes:

❑ Quality

❑ National or international standards

Quality
Medical and hygiene textiles should possess special qualities that ensure safety for the user. The product should be clean and not contaminated with dirt or bacteria. It should not contain chemicals which might be dangerous, such as artificial whitening agents, residues from bleaching, or insecticides used in growing. A special requirement may be good absorbency.

National or International Standards
International and national standards, such as British Pharmacopoeia (BP), contain detailed descriptions of construction and raw materials required for production. They also describe standard testing methods. Many countries require that medical textiles produced or used conform to standard specifications. Although standards vary according to country, most follow guidelines which ensure that basic quality is maintained. (See also Chapter 7.)

There are three basic medical textile products which can be produced on a small scale:

❑ Gauze

❑ Bandages

❑ Absorbent cotton

This chapter gives standard descriptions of and uses for these products, and a basic production method of manufacture. Further details, such as specifications for individual products and wet-processing methods are in the appendices.

GAUZE AND OPEN-WOVE BANDAGES

Gauze
Gauze is a lightweight fabric of open weave made from carded cotton yarn. Gauze should be soft, pliable and absorbent.

Most standard specifications require gauze to be bleached, clean and free from weaving defects, and contain no more than traces of seed coat or leaf, or other impurities. It should be free from chemical residues caused by wet processing. Artificial whitening agents should not be used in finishing the gauze because of the possibility of skin irritation. Gauze cloth should be packed in paper or plastic to keep it clean.

Gauze is most often used as an absorbent swab. It can also be used as a dressing, but this may cause problems if it sticks to the wound, risking disturbances to the healing process when removed. Gauze can also be combined with other textiles such as absorbent cotton to make absorbent pads (e.g. sanitary towels).

Open-wove bandages

Open-wove bandage consists of cotton cloth of plain weave. It is similar to gauze in structure. It is a porous, disposable strip of cotton fabric of one continuous length containing no joins, clean, and reasonably free from weaving defects. The cotton bandage is usually scoured and bleached, but can also be supplied unbleached and unsterilized, but always washed. Bandages are normally supplied in widths of 2cm, 3cm, 5cm and 7cm and in lengths of three to five metres.

Open-wove bandage cloth is most often used to protect dressings and hold them in place, and to give support.

Basic production process for gauze and open-wove bandages

A woven cloth structure which will not stretch in either direction is most suitable for making gauze or open wove bandages. Stretch bandages, however, can be made by the introduction of stretch yarns, by variations in cloth structure or by knitting.

The following step-by-step production process from the loom to the finished product assumes that a supply of suitable yarn is available.

The loom

Any type of simple two-shaft or four-shaft loom can be used to weave gauze or bandage. The width (reed width) of the loom will be determined by the width of the fabric to be woven. The normal widths of handwoven gauze and bandage are 36in. (90cm) or 48in. (120cm) therefore looms with reed widths of 40in. (100cm) or 52in. (130cm) should be used to allow for contraction of the cloth during weaving.

A handloom can produce gauze and bandage just as well as a powerloom. The obvious difference is that the powerloom is much faster, and is probably more economic to use.

The yarn

Yarn is available either direct from the spinner or from the market in hank, on cheese or on cone. If the yarn is purchased in hank, it will need to be wound on to convenient packages for warp preparation. The yarn is selected and purchased by specifying the quality (fibre content), thickness or count of yarn and the quantity by weight.

Warp preparation

Once the yarn has been purchased in a convenient package the first activity is to prepare a warp. There are several methods of warping for handlooms.

A warp can be made in a number of ways — with a single end on a simple wall-mounted warping frame; warping posts clamped to the top of a table; a line of sticks put into the ground; or an upright warping mill.

A warp can also be prepared from a number of ends which can be wound on to various devices mentioned above and also on to a horizontal warping mill. It is possible to make much longer warps on a horizontal warping mill. This is done by using the *sectional* warping method. The example below describes how many ends there are in one section and how many sections there are in the width of a warp (see Illustration 9).

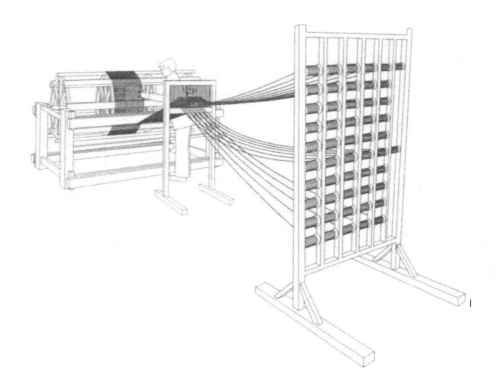

Illustration 9 Sectional warping

Before warping can start it is necessary to determine the *sett* of the cloth. The sett means the number of threads to a one centimetre or one inch width of warp and weft. The sett of gauze cloth is usually open and square. That is to say, the number of ends in one inch of warp is often equal to the number of picks of the same yarn in one inch width of weft.

If there are 20 ends per inch (epi) in a warp which is 40 inches wide, there will be $20 \times 40 = 800$ ends in the complete warp. In order give more strength to edges of the cloth during weaving it is sometimes necessary to introduce extra ends into the two selvedges. If the sectional warping method is used it is more practical to keep an even number of ends so that the warp can be divided equally. The warp containing 800 ends would be made up of 20 sections, each containing 40 ends. It will therefore be necessary to have a creel which can hold 40 spools of yarn, each spool containing one end of cotton yarn at least 100 yards in length. The length of the warp is determined by the length of the yarn on each spool. An example is given below.

Using 32s carded cotton, a warp for weaving 100 yards of gauze 40 inches wide can be prepared as follows (see Appendix 6 for summary of count system):

20 epi $\times$ 40 inches wide warp = 800 ends/20 sections each of 40 ends

The weight of 32s cotton yarn for this warp would therefore be
$800 \times 100/840 \times 32 = 2.98\text{lb}$ (say 3lb to include waste)

In this example the gauze would be square sett, i.e. same number of ppi (picks per inch) as epi, and the total weight of cotton required for 100 yards of gauze would be 6lb.

Using 40 spools, each holding sufficient yarn for 1 end $\times$ 20 sections ($1 \times 20 \times 100$ = 2000 yards $+$ 1 yard for waste) when using the sectional warping method, the warp is transferred from the warping mill on to the back beam of the loom. The individual warp ends are then selected in order from the *lease* and threaded into the *healds* supported by the shafts and drawn through the reed (probably two per dent). The small groups of about twenty ends each are finally secured on to the front beam so that all the ends have a uniform tension.

Weaving

The warp on the loom is now ready to be woven. If the loom has no take-up motion then the weaver will need to pay particular attention to the number of picks per inch (or picks per centimetre) being introduced across the warp and make sure that the number is constant. It would be appropriate to have 20 ppi to make a square sett cloth. However by reducing the number to 18 ppi, the structure would not suffer and the cost would be reduced accordingly.

Cloth finishing

Once the cloth has been woven and carefully taken off the loom it should, as far as possible, be kept rolled. The woven cloth should be kept as clean as possible. Before any further processing is carried out the cloth should be inspected and any knots should be taken out and any holes or weaving mistakes should mended.

Gauze and bandage is then wet processed in order to make the cloth clean, absorbent and white. Details of these processes are found in Appendix 3.

Rolling and packaging

Once the cloth has been pressed and dried the final steps of packaging take place. Methods for the rolling, packaging and labelling of gauze and bandages are described in Appendix 5.

ABSORBENT COTTON

Absorbent cotton, sometime referred to as surgical cotton or cotton wool, is an important medical textile in constant use, which requires no spinning, weaving or knitting.

Absorbent cotton is cotton fibre which has been cleaned, bleached and carded to produce a loose, absorbent wadding. It is possible to mix cotton with hollow fibre viscose which might produce a more economic product.

Absorbency is the main characteristic of this product. It should also be free from cotton leaf or seed and other foreign matter.

Absorbent cotton is used for cleansing and swabbing wounds, for pre-operative skin preparation, and for the application of medicaments to a wound. It can also be used as a pad, or as absorbent wadding for sanitary towels.

Basic production process for absorbent cotton

Absorbent cotton is produced from medium or short staple cotton, cotton waste after spinning or weaving, or a mixture of these with hollow fibre viscose. Producing absorbent cotton economically on a very small scale is difficult.

All cotton fibre, as in the preparation of cotton for spinning yarn, should first be opened and cleaned to remove all leaf, seed coat or other foreign matter. Illustration 1 on page 7 shows the blowroom section where the opening and cleaning takes place. The fibre is then scoured and bleached. Great care must be taken at this point in the production process to maintain the cleanliness of the fibre.

Following the opening and cleaning process, the bleached, dried fibre is carded to disentangle the fibre and produce a continuous, even web. Once the continuous web emerges from the carding machine, it is measured and rolled in one metre lengths which are cut into widths of 15cm or 30cm and packaged (see Illustration 10).

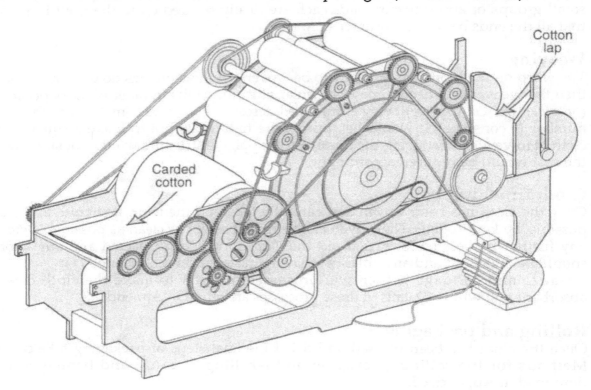

Illustration 10 Cotton carding machine

OTHER PRODUCTS

Other medical and hygiene textile products can be manufactured in a variety of ways using different processes and materials. These include different crêpe bandages, fabrics constructed from a combination of different fibres or yarns, and fabrics and combinations of fabrics and materials for different end-uses, and non-woven fabrics.

Domette bandage
Domette bandage consists of *union* fabric of plain weave, in which the warp threads are cotton and the weft threads are wool.

It is used for orthopaedic purposes, especially in cases where a high degree of warmth, protection and support is required.

Stretch bandages
Generally stretch bandages can be made by the introduction of specially twisted yarns, elasticated yarns, or by using a knitted structure. Often woven stretch bandages are referred to as crêpe bandages. They should be one continuous length containing no joins; clean, reasonably free from weaving defects, cotton leaf and cotton seed, and have edges which do not fray. Knitted stretch bandages are normally produced as tubes.

In the manufacture of woven crêpe, the warp is made from yarns which are twisted alternately S and Z. The crêpe effect occurs during the wet finishing process. After its removal from the loom, the fabric is washed in hot soapy water to relax the yarn, shrinking the fabric. It is then rinsed thoroughly and dried in its relaxed, unstretched state.

Cotton and wool crêpe bandage is also woven in plain weave. The warp yarns are alternately one end two-fold cotton and two ends singles woollen. The weft is singles cotton.

A crêpe fabric can also be woven by the introduction of an elastic yarn, such as core-spun rubber or elastomeric (i.e. Spanzelle or Lycra) cotton yarn.

Often a woven crêpe bandage will tend to fray at the edges. This fraying can be prevented by the introduction of a 2 or 4 end leno weave construction on each cut edge. Illustration 11 shows the basic construction of the leno weave and the dupes (special leno healds) on the shafts.

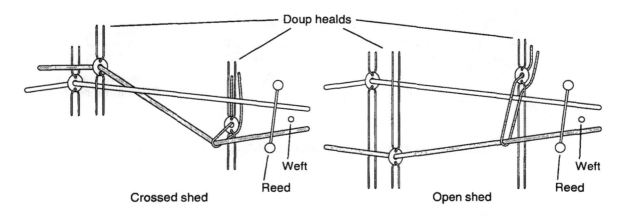

Illustration 11 Leno weave construction using special doup healds

Unlike plain open wove bandages, crêpe bandages will conform to the area of the body to which it is being applied. Crêpe bandages are used in the treatment of sprains and other conditions in which light support is required. They are also used as compression bandages.

An alternative method of producing a bandage is to knit it either by hand or on a simple knitting machine. Knitted bandages are, however, less common and would normally be mass produced in the form of knitted tubes. These knitted tubes can be knitted as thin as a finger or wide enough to pull over an arm or leg. Combed cotton yarns are usually used for knitted bandages.

There are several small-scale, electrically powered knitting machines on the market. The capital cost of these, however, is high and would not be suitable for use in remote rural areas without a constant electricity supply.

Nappies and diapers
The types of nappies or diapers vary greatly throughout the world, from reusable cloth to the modern disposable type, containing a high proportion of purified, absorbent fibres from wood pulp. The mass-production of the disposable type of nappy or diaper requires sophisticated machinery and therefore it is not within the scope of this handbook.

One of the more common reusable nappies is made from woven terry cloth. Terry cloth, sometimes referred to as turkish towelling, is a woven, warp-pile cotton fabric covered on both sides with uncut loops. The cloth can vary in thickness and weight according to the thickness and quality of the yarn used and the density of the structure.

Illustration 12 shows the cross-section and weave pattern of a standard 3-pick, double-sided terry cloth. Terry cloth can be hand-woven but it is necessary to prepare two warps on two back beams in order to produce this type of fabric. The ground warp is put on the lower beam and is woven under tension while the longer pile warp is on the upper beam. The loop is introduced into the woven cloth in a relaxed state at the fell of the cloth. The take-up, therefore, of the loop warp is much greater than the ground cloth and requires a warp which is sometimes four times longer than the ground warp.

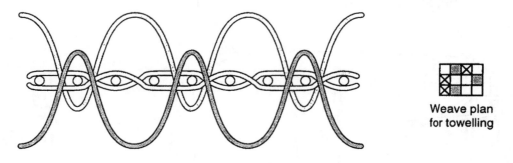

Weave plan
for towelling

Illustration 12 Construction of double-sided towelling

Sanitary towels
There are no standard specifications for producing sanitary towels on a small scale. Reusable sanitary towels tend to consist of a piece of absorbent cotton fabric. Plain woven cotton or terry cloth is appropriate for reusable sanitary towels.

A number of attempts have been made, with varying degrees of success, to produce disposable sanitary towels on a small scale. One possible method involves wrapping absorbent cotton wool with gauze to form a pad. A thin impermeable layer, such as polythene, is applied to one side of the cotton wool pad. The pad is then wrapped in a piece of gauze or introduced into a thin, open, knitted tube and stitched or knotted at both ends and packaged. Unless the basic raw materials — good quality clean absorbent cotton, gauze or knitted tube and polythene — are readily available, this method, which is also time-consuming, may not be economically viable.

4. TESTING AND QUALITY CONTROL

QUALITY CONTROL

It is clearly important that medical and hygiene textiles maintain a high level of quality, as their use affects the health of the user.

Within the process of small-scale medical textile manufacture there are a number of ways in which the quality of the end product can be controlled. The environment within which production takes place is important. If people work in adequate space and under reasonable conditions (i.e. lighting, temperature), for instance, the problems of quality control will be reduced. Similarly, if there are good tools and raw materials, poor quality products will be less likely to result.

One feature of textile manufacture is the constant presence of fibre fluff. This settles on all surfaces and when it becomes mixed with lubricating oils, it is frequently a source of oily stains on the cloth. Sweat and dust can also combine to make marks on the cloth. Although much of the grime will be removed during the wet treatment process, dirt and dust should be minimized during production. Each time equipment is empty, the opportunity should be taken to clear away fibre fluff. The entire work area should also be kept clean routinely.

TEST METHODS

Medical textiles are nearly always made to standard specifications. In many countries the distribution of medical textiles to hospitals and clinics is centrally controlled. Standards are strictly adhered to, and institutionalized testing is carried out on a regular basis. Even where standards are not systematically controlled, there will be a level of quality expected by the purchaser or user.

If national or international standards are being followed, testing for quality can be restricted to tests which check that the product conforms to that particular standard. If accurate testing has to be carried out to ensure that a textile product meets standard specifications, it is advisable to have samples tested by a fully equipped textile testing laboratory. Concerning sanitary towels made on a small scale, the procedures for testing woven and knitted fabrics and absorbent cotton should be followed. There are also simple test methods which ensure that consistent product characteristics are maintained.

Absorbent cotton
Absorbent cotton comes in a variety of forms, depending on its origin and its intended use. Absorbent cotton should have a number of standard characteristics, outlined in Chapter 3.

Yarn
Supplies of the correct yarn should be readily available and of consistent quality. Unevenness in the yarn might be due to the quality of the cotton fibre used. Yarn should be checked regularly to ensure that it has the correct count. A simple twist test can be achieved using the twist testers shown in Illustrations 13a and b. These machines hold one end of the yarn firm and the other in a jaw which will rotate so

untwisting the yarn. At the same time the number of twists in a consistent length of yarn (say 10 inches/25cm) is recorded. It is important that the yarn is tested under constant tension. A minimum of ten tests should be carried out at regular intervals along the yarn.

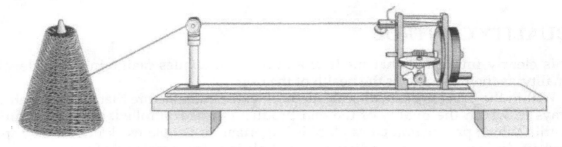

Illustration 13a Commercial twist testing apparatus

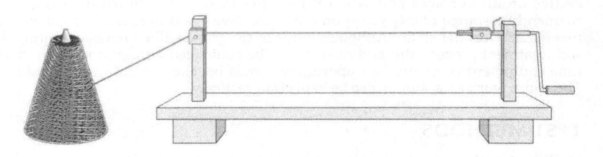

Illustration 13b Simple twist tester

Fabric construction

Simple fabric tests include measurement of the number of threads per square unit, dimensions and weight of fabric.

Woven fabrics

Number of threads per unit square. One method of checking ends and picks per centimetre (cm) is to use a frame as shown in Illustration 14. This is marked in cm and mm, and is placed on the fabric. The number of ends and picks within the frame are counted. If the threads cannot be easily seen individually, a pick glass or linen prover is used as well, shown in Illustration 15. This known as checking the sett of the cloth.

Weight per unit area. This test is carried out by taking several samples of standard size, such as one metre square, weighing each and then taking the average figure. A simple weighing scale is needed for this test.

Width. Consistency of weaving depends on regular pick spacing, constant warp tension, even weft tension and evenly wound pirns. This will help to produce a standard width fabric, particularly when a temple is used. The fabric is then laid out on a flat surface and a ruler or tape measure used to determine the average width by taking several measurements at intervals along the fabric.

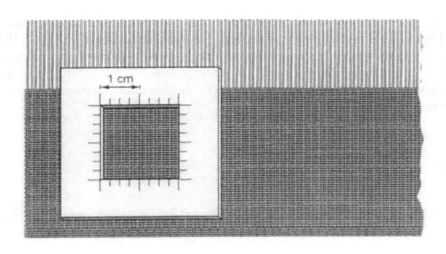

Illustration 14 Frame to determine the sett of cloth

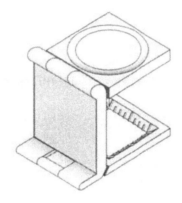

Illustration 15 Pick glass or linen prover

Knitted fabrics

Even quality is achieved by the use of high quality yarn, while maintaining consistent course length and loop length.

The course length is the length of yarn knitted into one course of the fabric and is a good indicator of quality. Course length can be determined by unravelling one course from the fabric. The yarn is held straight (without kinks) and its length measured.

A method of checking loop length during knitting is to measure one metre of yarn between the knitting machine and the cone feeding it, and to mark each end of this metre. The yarn is then knitted into the fabric, and the number of stitches (n) knitted from the metre is counted. This will give a value for the loop length (L), from ($L = 100/n$ cm).

STANDARD TEST METHODS

Other tests ensure that medical textile products conform to standard specifications. The methods vary in complexity, and many require the use of a fully equipped testing laboratory. The methods described below are simplified testing procedures found in most national and international standards. While useful for controlling the consistency and quality of the products being produced, these methods do not by themselves ensure standard quality.

Determining absorbency (for absorbents and dressings). Absorbency is usually measured by selecting three dry samples of consistent size or weight. Illustration 16 shows the test procedure for absorbency. Each sample is first compressed in a small jar and then introduced into a flask of water. The sample is timed to become fully saturated and sink to the bottom of the flask. A glass flask and distilled or de-ionized water are needed for this test.

Illustration 16a Sample compressed in small jar

Illustration 16b Sample is introduced into a flask of water

Illustration 16c Sample is timed to become fully saturated and sink to the bottom

The initial absorbency of a product is diminished after long storage (two years or more) or exposure to heat and damp. Repeated sterilization by autoclaving is also likely to impair absorbency.

Fluorescence (for bleached products). This test is to ensure that fluorescent brightening agents are not used during wet processing. When a sample is examined under screened ultra-violet light, not more than an occasional point of fluorescence should be visible.

Acidity or alkalinity. This test is to ensure that the material has a neutral pH value. A sample of material (10g) is soaked for half an hour in 100ml cold distilled water. Decant the extract, carefully squeezing out the residual liquid from the sample. The liquid is tested with pH paper, as seen in Illustration 17. The pH paper changes colour to indicate acidity or alkalinity. The pH value should be neutral.

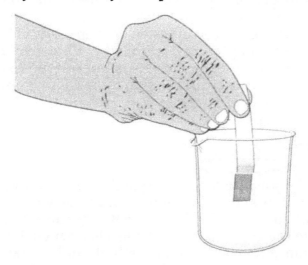

Illustration 17 Testing for acidity or alkalinity

Freedom from detergents. To ensure that no residual detergent is left in the material following wet processing, a small sample is cut into pieces and boiled in 200ml water for half an hour. The water is decanted into a 500ml flask. This process is repeated twice with the same sample. This flask of aqueous extract, when gently shaken, shows no appreciable signs of frothing. A heat-proof flask, water, and a stove or burner are needed for this test.

Freedom from starch. This test ensures that all starch-based size has been removed from the fabric. A small sample of fabric is thoroughly wetted with warm water and then a drop of iodine or potassium iodide solution is spotted on the fabric. A strong blue colour indicates the presence of starch. If the spot remains pale yellow, all starch has been removed.

Sterilization. The tests for sterilization require complex laboratory procedures, using sophisticated equipment, and are beyond the scope of this handbook. For more information on sterilization see Appendix 4.

5. PLANNING FOR PRODUCTION

The circumstances in which medical textiles are to be made are likely to vary widely. It may be intended to establish production in a new situation, where there is little local textile experience. On the other hand, perhaps there is an established local activity with considerable expertise in textile or even medical textile production, and this is to be expanded, adapted or improved. Whatever the situation, careful planning is important, and a few basic considerations may help to achieve a successful result. This is particularly important if commercial production is being planned.

MARKET CONSIDERATIONS

It is essential to have a solid understanding of the needs and markets for the product before manufacture is planned. In many countries the distribution system for bandages, absorbent cotton, and gauze is centrally controlled. The purchase and supply often operate through a centrally controlled national or international tender system. Competition for tenders might be high and it may be difficult for new producers to enter the market. There may also be a government-regulated buying price. It is important therefore to be aware of the size and nature of the competition.

A market survey should obtain the following information:

❑ What and where are the institutional (hospitals, clinics, pharmacies) and household markets?

❑ Which products are currently being used, and which are needed?

❑ What are the quantities currently supplied to the market, and what is the estimated demand?

❑ Does supply meet demand?

❑ What are the distribution channels from producer to consumer, the means of payment, and the distribution costs?

❑ What are the existing wholesale and retail prices of the product? If imported, what are the import prices and duty costs?

Medical textiles differ from other textile products, in that there is often little scope for diversification and design variation, which would give it a competitive edge in the market. It is therefore important that the items under consideration can be produced and sold at a competitive price.

If the production of sanitary towels or nappies is being considered, there should be a clear identification of local needs and markets, and an understanding of the cultural, social and economic context in which these items are purchased and used. Guidelines for investigating the social context of sanitary protection practices are in Appendix 1.

PRODUCTION CONSIDERATIONS

It is important to know what the national or international standards are, how these are regulated, and what the procedures are. Copies of standard specifications and procedures for registration are usually available from appropriate government departments.

The space in which production takes place, be it a household or a central location, needs to be considered. It should be dry, light and airy with good shade to promote a good working environment. The equipment and space should be kept as clean as possible. Storage of yarns, chemicals and other raw materials should be away from direct sunlight. All materials should be stored off the ground on wooden shelves or slats, and should be checked frequently for signs of attack by insects or mildew.

There should be careful consideration of where and how the wet processing is to take place. An adequate supply of good quality clean water is an important requirement, for de-sizing, scouring, bleaching, and washing the finished product. If a heat source is needed for scouring and bleaching, then there must be a cheap and consistent supply of fuel. If production is at individual household level and space is limited, a centralized wet processing unit, shared by a number of producers, could be organized. Ideally there should be an adequate space for drying, which is clean, dry and well-ventilated. If weather permits, products could be dried in the sun. There should be additional space for inspection, testing, packaging, and storage for convenient delivery.

Training is an essential step if the equipment or techniques being used are new to the area. If possible, some training should be started, even if for only one or two key people, before the establishment of a new unit. Suitable training is sometimes available at local textile centres or educational institutions, so these possibilities should be investigated. Training programmes should also include the safety aspects of using new and unfamiliar equipment.

Before making any decisions about the scale and type of equipment needed in a new situation, the following points should be carefully considered.

Cost Is there sufficient justification or need for the level of expenditure planned? Is there sufficient cash available to meet purchasing costs? This includes not only the cost of the basic equipment, but a stock of spare parts, ancillary equipment, wages, day-to-day running expenses and raw material stocks.

Capacity Does the choice of equipment used match the desired production of medical textiles and the available raw material supplies? Is there room for expansion?

Location If all the equipment is located in one place, will this suit all those who are expected to use it? Would temporary locations and portable equipment be more suitable? If materials have to be moved between different locations, is suitable transport available at a reasonable cost?

Type Is the equipment suitable and flexible enough for medical textile production? Can it make other products which will allow for changes in the market or seasonal variations? Will the equipment stand up to daily use, and how long is it expected to last?

Availability Is the equipment available locally from a reputable manufacturer? Are specifications available from the manufacture of equipment? Are there sufficient skills to carry out the construction? What kind of spare parts and maintenance services are available? If power-driven equipment is being planned, is a reliable power supply available for a reasonable period each day?

Experience Is the equipment easy to use? Can tuition be obtained locally, or at a national training centre? What local skills exist, or could be learned through training, for maintaining and servicing the equipment?

Social acceptability Will the choice of equipment and system be readily acceptable to people? Is the way in which production is organized acceptable to the community? What changes to existing social practice will be required? Has the demand for change come from the local community or from outside? What impact would new or changed production have on other existing and similar producers in the area?

Benefits If existing production is being changed or improved, how will the producers benefit? How will this be made clear to them?

Expert advice Seek expert advice about the equipment and how you propose to use it before purchase. Shortlist the most suitable range of equipment, evaluate it financially and technically to help identify the best final choices.

FINANCIAL FEASIBILITY

Once the operation has been planned in some detail, it is useful to undertake a trial costing for the production of medical textiles, to give a rough idea of the viability of the operation in commercial terms. The cost of production generally has two components: fixed or indirect costs, and variable or direct costs.

Indirect costs

- ❑ Interest on the cost of any stock of raw materials
- ❑ Cost of premises
- ❑ Heat/light/power
- ❑ Cost of depreciation of equipment and interest on any loans for purchase
- ❑ Consumable materials
- ❑ Administration costs
- ❑ Marketing costs

Direct costs

- ❑ Raw materials
- ❑ Wastage
- ❑ Transport
- ❑ Wages (including any contribution to a welfare fund and any incentive wages)

Break-even analysis To estimate the production output needed to compete in the market, first of all assess its financial feasibility. A break-even analysis considers the following questions.

❏ At what level of production will the product be able to cover its direct and indirect costs?

❏ What is the minimum price needed for the project to be able to cover its fixed and variable costs?

❏ What is the minimum price needed for the project to be viable at different levels of production?

❏ What happens if financial assumptions of costs or prices are changed?

❏ What are the best, worst, and probable scenarios for the project?

The estimated production output needed to break even should be compared with the capacity of the equipment and with the size of the expected market. To check the viability of the planned unit, the cost of production should also be compared with prices in the market.

6. EQUIPMENT SUPPLIERS

HAND LOOMS

United Kingdom
AVL Looms Ltd,
St George's Mill, St George's Street,
Macclesfield, Cheshire
(also in USA: 601 Orange Street,
Chico, California 95206)

Emmerich (Berlon) Ltd,
Wotton Road,
Ashford, Kent TN23 2JY

Bonas Griffith Ltd,
12a Southwick Industrial Estate,
Sunderland SR5 3TX

Sweden
Berga Hemslojdens Ullspinneri AB,
S-783 02 Storå, Skedvi

India
V.P.F. Testing Equipment,
Ventakapathy Foundry,
Pellameedu, Coimbatore 641004,
Tamil Nadu

M/S Kamal Metal Industries,
Gajjar House,
Astodia Road,
Ahmadabad 380001,
Gujarat

KNITTING MACHINES

Europe
Textilmachinenfabrik Harry Lucas GmbH & Co.,
Postfach 2003,
D-2305 Neumünster,
Germany

Edouard Dubied & Cie S. A.
(Dubied hand and power knitting machines)
CH-2001 Neuchâtel,
Switzerland

Semexco di H. Gierlinger & C s.n.c., 20090
Trezzano s/n (Milano) Via Morona 79,
Italy

Frame Knitting Ltd,
PO Box 21,
Oakham,
Leicestershire LE15 6XB,
UK

Jones & Brother,
Knitting Machine Division,
Shepley Street, Audenshaw,
Manchester M34 5JD,
UK

PORTABLE AUTOCLAVES (steam sterilizers)

Europe
Phillip Harris Scientific,
618 Western Avenue, Park Royal
London W3 OTE,
UK

Kelomat Gruber & Kaja KG,
Obere Dorfstrasse 1, A-4050 Truan,
Austria

India
J.T. Jagthiani,
National House, Tulloch Road, Apollo Bunder,
Bombay

Repute Scientific Co,
13/21 3rd Panjarapole Lane,
2nd Floor, SVU Bg CP Tank Road,
Bombay 400 004

TESTING EQUIPMENT

India
V.P.F. Testing Equipment,
Ventakapathy Foundry,
Peelameedu, Coimbatore 641004,
Tamil Nadu

M/S Kamal Metal Industries,
Gajjar House, Astodia Road,
Ahmadabad 380001,
Gujarat

7. SOURCES OF FURTHER INFORMATION

RESEARCH ORGANIZATIONS

❑ Information services on a wide range of textile topics are usually available within research organizations. This usually includes book lists and copies of research papers.

❑ They will usually undertake testing or other work for which a charge is made.

❑ They sometimes organize courses or training programmes on aspects of textile manufacture.

United Kingdom

British Textile Technology Group (BTTG),
Shirley Towers, Didsbury,
Manchester M20 8RX
Tel. 061 445 8141
Telex 668417
Fax 061 434 9957

British Textile Technology Group (BTTG),
WIRA House, West Park Ring Road,
Leeds LS16 6QL
Tel. 0532 781 381
Telex 557 189
Fax 0532 304 195

British Textile Technology Group (BTTG),
Newton Business Park, Talbot Road,
Hyde SK14 4UQ
Tel. 061 367 9030
Fax 061 367 8845

Dr Stephan Thomas,
Surgical Materials Testing Laboratory,
Bridgend General Hospital,
Corlea Road,
Bridgend CF31 1JP
Wales

India

Ahmadabad Textile Industry Research Association (ATIRA),
Polytechnic Post Office,
Ahmadabad 380015,
Gujarat

Textile and Allied Research Association (TAIRO),
Baroda,
Gujarat

PUBLISHED SOURCES OF INFORMATION

Textiles

The following books contain basic information on textile production and testing.

Hall, A.J., *Standard Handbook of Textiles* (Butterworth and Co.)

Emery, I., *The Primary Structure of Fabrics* (Washington DC)

Marks R. and Robinson A.T.C., *Principles of Weaving* (Textile Institute)

Spencer D.J. *Knitting Technology* (Pergamon Press)

Booth, J.E. *Principles of Textile Testing* (National Trade Press)

Foulds, J., *Dyeing and Printing*: *A handbook* (Intermediate Technology Publications)

Foulds, J., *Spinning*: *A handbook* (Intermediate Technology Publications)

Iredale, J., *Yarn Preparation*: *A handbook* (Intermediate Technology Publications)

Newton, A., *Fabric Manufacture*: *A handbook* (Intermediate Technology Publications)

Medical textiles

Thomas, Dr S., *Wound Management and Dressings* (The Pharmaceutical Press)

The Journal of Wound Care (Macmillan Magazines Ltd) Basingstoke RG21 2XS, UK

Detailed specifications and simple methods of tests can be found in various pharmacopoeia, such as the British Pharmacopoeia, or European Pharmacopoeia, which are available in reference libraries. National standards can be obtained from most government standards testing institutes.

Feasibility studies and marketing

Feasibility Studies: Training activities and guidelines to determine if a business is a good idea (OEF International): 2101 L Street, NW, Suite 916, Washington DC 20037, USA

Marketing Strategy: Training activities for entrepreneurs (OEF International)

APPENDIX 1

CULTURAL ASPECTS OF SANITARY PROTECTION PRACTICES

In many countries the social and cultural issues surrounding sanitary protection practices are very complex. The materials a woman uses depends on the surrounding culture, her social and economic status, and what is available locally. A thorough investigation of existing sanitary protection practices should be undertaken before decisions about production are made. Information should be obtained about the existing market, and on the social, economic and cultural constraints surrounding use and availability. Guidelines for market research are described in Chapter 5. The following questions may provide useful guidelines for obtaining social and cultural information:

❑ What materials are used during menstruation, including commercial sanitary towels, old clothing, etc., and how do women care for (wash, dry, store) or dispose of these materials?

❑ What are the principal problems associated with menstruation, in terms of health and cultural constraints?

❑ What are the social and cultural limitations dictating the use, availability, and access to sanitary protection materials? Briefly outline the decision-making structure within a household with respect to the purchase of materials.

❑ What problems are caused by the use and care of sanitary protection materials?

❑ How do women cope with the problems identified above, and what solutions do they suggest to address these problems?

❑ Are commercial sanitary towels (napkins) accessible and affordable, and are women willing to purchase them?

Any investigation into sanitary protection practices should be conducted carefully and with sensitivity.

Focus group discussion
The focus group discussion is a useful method for data collection. It involves bringing together a small group of 8-10 women of similar social and economic backgrounds for an informal discussion. A set of questions can be used as guidelines, but should not limit or discourage participative discussion. A number of focus group discussions and other participative methods of data collection can provide useful insights into sanitary protection needs.

APPENDIX 2

SPECIFICATIONS FOR MANUFACTURE

The following specifications for gauze and open-wove bandages are based on the national standards for Bangladesh, Sri Lanka, some African countries and the United Kingdom. They are intended as a guide only and may be modified according to local conditions.

Gauze

Yarn specification

warp	unbleached singles carded cotton (unsized); 18 Tex (32s c.c.)
weft	as above

Construction

warp	8 ends per cm (20 epi)
weft	6 picks per cm (16 ppi)
weight	20g per m² (0.6oz per yd²)
finishing	scour, bleach, rinse, dry thoroughly and package

White open-wove bandages

Yarn specification

warp	unbleached singles carded cotton (sized); 16 Tex (36s c.c.)
weft	as above

Construction

warp	16 ends per cm (40 epi)
weft	10 picks per cm (25 epi)
weight	36g per m² (1oz per yd²)
finishing	scour, bleach, rinse, dry thoroughly, cut to required widths and lengths, package

Absorbent cotton

staple length	not less than 10mm
absorbency	average sinking time should not exceed 10 seconds.

APPENDIX 3

WET PROCESSING

The wet processing stages of medical and hygiene textile production include de-sizing, scouring and bleaching. These processes bring about the necessary characteristics of absorbency and whiteness.

There are a number of different chemicals and processes that can be used for wet processing, depending on the availability of materials and equipment.

De-sizing
Some fabrics may contain sizes, which are added to reduce friction and breakage during weaving. If not removed thoroughly size will prevent proper and even scouring and bleaching, and will lead to a material that is stiff and rough-textured. Most standards require medical textiles to be free of all traces of sizing agents.

Scouring and bleaching
Scouring is an alkaline process which aims to aid the breakdown of the cotton seed residue and the removal of fats and waxes. Bleaching destroys the coloured substances and adds an even whiteness. Both processes are necessary to make the fibre highly absorbent.

Rinsing
All wet processes require a thorough rinsing of the material after a chemical treatment, in order to remove all of the chemicals, which have been strongly absorbed by the fibre. Medical textiles should be as free as possible from chemical residue, and should be pH neutral; that is, there should be no acid or alkaline reaction when tested. It is important that there is a reliable supply of clean water for thorough rinsing after each of the wet processing stages.

Drying
Medical textiles should be dried in clean conditions. Most products can be sun- or air-dried. After drying, the products should be handled as little as possible, to reduce the risk of unnecessary contact with germs.

GENERAL METHODS OF SCOURING AND BLEACHING COTTON

Scouring and bleaching cotton fibre, yarn and fabric is usually carried out either as a two-stage process or a one-stage process. Both are described below.

Two-stage process
This process uses hypochlorite as the bleaching agent (either bleaching powder or sodium hypochlorite solution). The reason for the two-stage process is that hypochlorites must be used cold in strongly alkaline conditions, if the fibre is not to be chemically damaged. The removal of impurities such as wax, however, requires hot treatment. This process is extremely effective when properly used, giving a very thorough purification of cotton.

Method

Scour the yarn or loose cotton in a suitable container, in solution containing 2g/litre Lissapol N, 2g/litre of sodium carbonate and 4g/litre of sodium hydroxide, by heating at as high a temperature as possible for 30 minutes (boil if possible).

Wash in cold water.

Bleach the yarn or loose cotton in a suitable tub. The bleaching solution should contain 1–3g/litre of chlorine, and 5g/litre of sodium carbonate. The pH of the solution should not fall below 9–10 during bleaching. The material should soak at 20°C for several hours (or overnight), or until the cotton is sufficiently white.

Wash thoroughly in cold and then hot water.

Acidify in a solution containing 1–2g/litre hydrochloric acid for 20 minutes at 20°C, and then wash thoroughly in cold water. This treatment neutralizes the residual alkali in the bleach.

One-stage process

This process uses hydrogen peroxide as the bleaching agent. If peroxide is available, scouring and bleaching can be carried out simultaneously. Equipment must be made of appropriate materials mentioned in Note 1 (page 39). The water supply also needs to be free from any traces of metals or colour (e.g. water with a high iron content should not be used).

Method

Scour and bleach the cotton yarn or loose cotton in a solution containing 7g/litre sodium silicate (79°Tw sol.), 0.5–1g/litre sodium hydroxide, 2g/litre sodium carbonate, 0.5g/litre Lissapol N, 7.5ml/litre hydrogen peroxide (35% sol.). Treat the cotton in this solution at 30°C for 20 minutes and then gradually raise the temperature to 80–90°C and maintain this temperature until sufficiently bleached, moving the yarn or stirring the loose material to ensure even treatment.

Wash thoroughly in hot then cold water.

Acidify as in the two-stage process.

SCOURING AND BLEACHING FABRIC USING A WINCH

Short lengths of gauze fabric can be bleached in vats as above. Because it is difficult to handle long lengths of wet fabric, it may be necessary to have more complex equipment. A winch is most useful for this process (see Illustration 18).

The winch shown in Illustration 18 is useful for lengths of fabric up to 50 metres. The fabric is sewn on to the machine in a continuous length either in the form of a rope, with the fabric bunched together, or with the fabric held at full width. Roping the fabric is most suitable for lightweight and knitted fabrics.

Two-stage process using hypochlorite

The conditions and concentration of reagents are the same as for yarn scouring and bleaching. De-size by steeping in a pit with either water or enzyme. Wash on winch. Scour on winch and then bleach in a pit with hypochlorite. Wash, acidify, and rinse on winch. Dry.

One-stage process using peroxide on winch

Scour and bleach the fabric in a solution containing 7g/litre sodium silicate (79°Tw sol.), 0.5–1g/litre sodium hydroxide, 2g/litre sodium carbonate, 0.5ml/litre Lissapol N, 7.5ml/litre hydrogen peroxide (35% sol.).

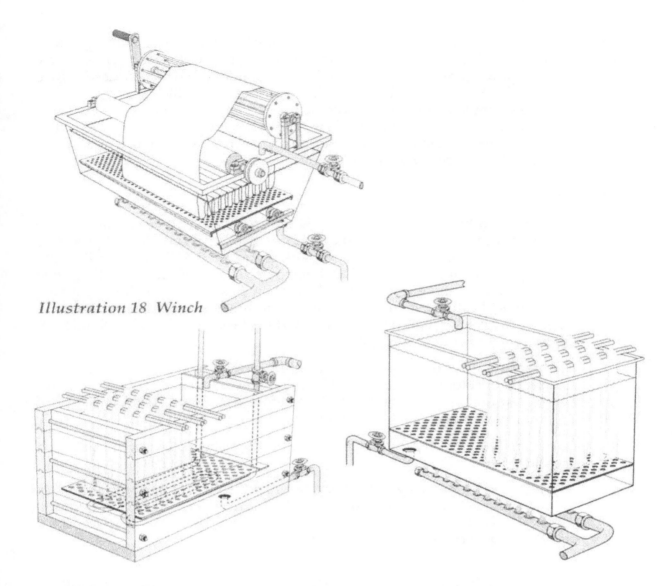

Illustration 18 Winch

Illustration 19 Scouring or bleaching
using cold water processes or indirect
heating in wooden vat

Illustration 20 Scouring or bleaching
using direct heating in metal vat

Raise the temperature to 50°C and run the fabric for 15 minutes, then slowly raise the temperature to a boil, or as near as possible, for 30 minutes. Continue to run at this temperature until bleaching is complete (about one hour). Wash very thoroughly in warm and then cold water. Acidify in a solution of 1–3g/litre hydrochloric acid at 20°C for 20 minutes, then wash in cold water.

DE-SIZING FABRIC USING SIMPLE EQUIPMENT

Steeping method The fabric may be steeped in warm water (35–40°C) in a concrete or tile lined tank and left overnight.

To speed up the steeping process enzyme preparations may be added to the water which are specific for the breakdown of starches. These speed up the removal of starch but must be used under exactly the correct conditions for the particular enzyme being used (temperature and pH), since they are active enzymes, and can be 'killed' quite easily. They are generally used at concentrations of 0.5–1g/litre, under conditions specified by the manufacturer.

Oxidative de-sizing Many methods are possible, using acids, bleaching agents or other chemicals. This process has the advantage that it will remove most sizes. However, it requires careful control to avoid fibre damage, and is best used where appropriate equipment is available, and accurate control is possible.

One relatively simple method which can be used if only a steeping pit is available involves saturating the sized cotton fabric with a solution containing 2–3g/litre chlorine and 5g/litre sodium carbonate. Steep in a covered pit for four hours. Wash very thoroughly before scouring and bleaching.

NOTES

1.   Scouring and bleaching can be carried out in any container of suitable size made from concrete, glass, ceramic, stainless steel, or wood. If cold processes or indirect heating is used, a wooden tank represented in Illustration 19 could be used. If the process requires high temperatures using direct heating, then stainless steel is the best material. Illustration 20 shows a stainless steel tank using direct heating. Equipment made from iron, copper, brass, lead, nickel, or any alloy containing these metals should not be used. Bleaching should not be carried out in direct sunlight.

2.   Both sodium hypochlorite and bleaching powder are sold with a concentration of chlorine content. As bleaching powder loses its strength on prolonged storage, it is advisable to test before use. If it smells strongly of chlorine, the strength is usually good.

3.   Yarn can be hung on a line to dry. Cotton fibre should be spread out on a clean even surface to dry.

4.   The handling of long lengths of fabric can be difficult. This can be made easier by using a handcart for transporting the fabric from one tub to another. Alternatively, a series of pulleys attached overhead can be used to transport fabric in rope form (see Illustration 21).

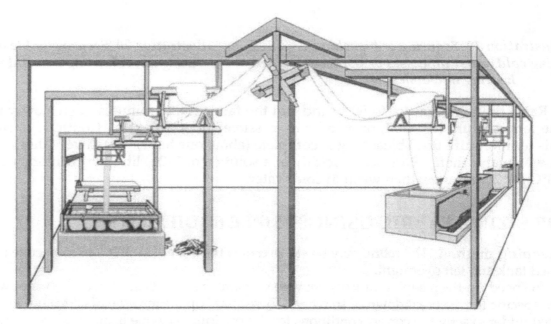

Illustration 21 Small, manually operated bleaching unit

5.   When using a winch, it is very important that it is made of suitable materials. High quality stainless steel is best, but iron or steel can be coated with silicate or cement. The bushes and bearing for the guide rollers, which are under the surface of the bleaching solution must also be of suitable materials (carbon or nylon).

6.   Fabric can be hung on a simple rack to dry (see Illustration 22).

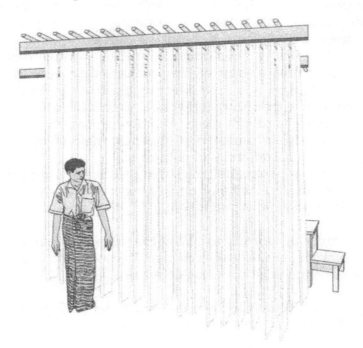

Illustration 22 Simple rack for drying cotton gauze

EFFLUENT DISPOSAL

Although the toxicity of the effluent from de-sizing, scouring and bleaching activities is not likely to be a problem on a small scale, there should be some consideration of drainage, disposal or treatment of effluent. The quantity of effluent needing treatment or disposal is affected by the size of the activity and the increase in production.

Ideally, the treated effluent should be colourless and have a neutral pH. It should be capable of being diluted in a large water system without causing any problems. Carrying this out would represent an additional cost, in terms of both increased overheads and increased running costs. There is a range of simple treatments and methods of disposal which can be used in particular circumstances.

The simplest and cheapest method is to run the effluent directly into a moderately large pond (say 10 metres by 10 metres) which is used solely for the purpose of its storage and dilution. This should minimize problems with drainage into ground water. A combination of sedimentation and biological action will take place over a period of time, particularly in the presence of strong sunlight. This method will break down the chemical residues, and will cause a deposit of sludge to form. Fermentation problems may occur in hot weather, but this is not likely to happen for a long time unless the pond is too small.

APPENDIX 4

STERILIZATION

Sterilization is the process of killing or removing germs. A medical textile product should be sterile depending on its use. For example, if an item is in direct contact with a wound or is used in operations, it should be sterile.

Many hospitals and clinics in developing countries sterilize their own medical textile supplies, so sterilization as part of the production process does not normally take place. It is important, however, that after wet processing, products are handled as little as possible, to reduce the risk of unnecessary contamination.

If sterilization is required as part of the production process, it can be carried out in a number of ways. The easiest, safest, and cheapest method in small-scale manufacture is by using steam sterilization in an autoclave.

Sterilization always takes place after packaging. It is important that appropriate packaging materials are used. Choice of packaging materials is determined by the following considerations (see also Appendix 5).

❑ The package must be porous to allow the steam to reach the article to be sterilized. Thick paper such as cartridge (heavy white drawing paper) and cloth are therefore the most appropriate materials.

❑ The packaging must be able to keep the article sterile. Paper which tears easily and packages with weak seals must be avoided.

❑ The article must be labelled, so that the user knows that the product has been sterilized.

Record keeping
It is important to know that an article has been sterilized. A simple label saying 'sterile' may not be enough to show if the product is sterile. Packages could for instance have heat-sensitive markers on the labels which indicate whether or not they have been sterilized. Batches should be stamped and recorded so that they can be traced back to a particular process on a particular day.

Autoclaving
An autoclave is an airtight metal vessel in which steam can generate pressure, raising the temperature to 121–134°C. On reaching the appropriate temperature, the items are subjected to a saturated steam environment for a specified period of time. For temperatures and timing see autoclave manufacturers' instructions.

There are a range of sizes and types of autoclave available (see Chapter 6). The autoclave used should be suitable for steaming textiles. It should be capable of drying the product following steam treatment. This can be achieved by equipping the autoclave with a valve which releases steam following sterilization. The autoclave should also be equipped with a thermometer and a pressure gauge so that the appropriate pressure can be accurately maintained. These must be checked regularly to see that they are working properly. Operating instructions must be followed carefully. Illustration 23 shows an example of a small, simple autoclave.

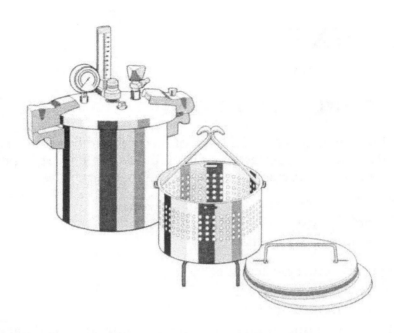

Illustration 23 Simple autoclave

The essential parts of the steam sterilization process are:

1. Pack the articles loosely in the autoclave to allow the steam to penetrate all the items thoroughly.

2. Replace as much air as possible inside the autoclave with steam. The presence of air pockets might prevent the steam from penetrating all of the items. In larger capacity autoclaves, air is extracted with a vacuum pump, and replaced with steam. If the autoclave has no vacuum pump, allow the steam to flow through the load for several minutes before bringing the chamber up to pressure.

3. Bring the chamber up to pressure, check temperature, and hold the pressure and temperature steady for a specified length of time.

4. Release the steam through the valve and let in air, or open the chamber and let the contents dry naturally.

5. Autoclave manufacturers' instructions should be read and followed carefully to avoid accidents.

APPENDIX 5

PACKAGING AND LABELLING

Packaging and labelling are essential parts of medical and hygiene textile production. There is a wide variety of materials and methods for packaging and labelling a product, and the costs and benefits should be weighed carefully.

Benefits of packaging
Packaging has a number of benefits:

- ❑ keeping the product clean and dry
- ❑ keeping the product compact
- ❑ keeping the product sterile (if sterilized)
- ❑ protecting the product from tampering or adulteration and damage
- ❑ facilitating marketing and selling
- ❑ conforming to a standard.

Packaging materials
The material used for packaging should not adversely affect the product. For example, waxed paper should not be used in direct contact with materials such as absorbent cotton and gauze, as it reduces absorbency. The most common packaging materials used for medical and hygiene textiles are paper and polythene. Other considerations for choosing types of packaging materials may be:

- ❑ customer type
- ❑ customer needs
- ❑ style of presentation
- ❑ brand image
- ❑ strength
- ❑ reusability
- ❑ resealability
- ❑ suitability for sterilizing (see Appendix 4)
- ❑ packaging equipment needed
- ❑ availability
- ❑ cost.

Benefits of labelling
Labels have a number of useful benefits. They inform the user of the content of the package. They also facilitate the marketing of the product, by promoting a brand image.

A label for a medical textile product should contain the following information:
- ❑ the name of the product
- ❑ manufacturer's name, trade mark and address

- □ volume/weight/quantity of contents
- □ shelf-life and lot number
- □ indication of sterilization
- □ material content
- □ instructions for use
- □ standard certification mark (with permission of the relevant standards testing institution).

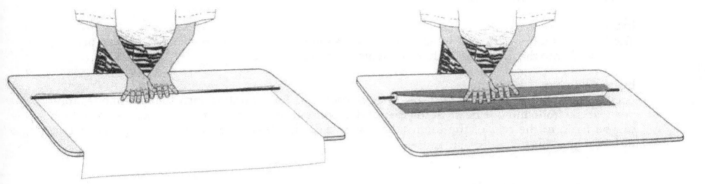

Illustration 24a Bandage fabric is rolled on to metal rod

Illustration 24b Rolled bandage is wrapped with paper and rod removed

Methods

Bandages and gauze can be wrapped in bulk or divided into smaller convenient pieces and individually wrapped. Bandages need to be cut into convenient sized rolls, usually 3 metres long, and wrapped individually. One method for producing individual bandages is as follows.

The full width of the cloth is rolled on to a metal rod with a diameter of about 1 centimetre. When the desired length of fabric (e.g. 3 metres) is rolled, the roll is wrapped in paper, glued and the metal rod is removed (see Illustrations 24a and b). The roll is then cut accurately into carefully measured widths. Cutting can be done with a circular, rotating blade (see Illustration 25). Care must be taken to protect the circular, rotating blade. The individual cut rolls are then made up into packages of ten or twelve bandages.

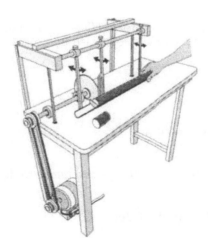

Illustration 25 Simple bandage cutting machine

APPENDIX 6

GLOSSARY

bar (barry or barriness)
A mark in the form of a bar across the full width of a piece of woven cloth which differs in appearance from the rest of the cloth. Often a mistake in weaving caused by either incorrect picking, or wrong weft or yarn tension.

batten
The frame containing the *reed* which the weaver swings to and away from himself when beating up the weft into the *fell* of the cloth. Known also as a *sley* or beater.

beam
A cylinder of wood or metal with bearings at each end for mounting into suitable flanges, one beam at the front and one beam at the rear of the loom. A double beam refers to two beams which can be fixed to the rear of the loom when two warps are taken up in the weaving under two different tensions.

beer
A group of 40 warp threads.

c.c.
Abbreviation for English cotton count system for yarn.

count
Systems of numbering, for identification, and methods of calculating the sizes of yarn either by weight per unit length (direct fixed-length system used for filament yarn counts), or by length per unit weight (indirect fixed-weight system used for cotton or woollen counts).

counting glass
A magnifying glass mounted in a small hinged metal frame with a fixed focus, the base having an aperture measuring either one square inch or one square centimetre. Used for counting the *ends* and *picks*, *courses* and *wales* in a fabric. Also known as linen prover or pick glass (see page 24).

course
A row of loops across the width of a flat knitted fabric or around the circumference of a circular fabric.

cover
The evenness of yarn spacing giving a uniform effect on the surface of the cloth. The degree to which the underlying structure of a fabric is concealed by finishing.

dent
Spaces between the *reed* wires.

doup
Half *heddle* used in leno or gauze weaving.

drafting
The process of drawing out *laps*, *slivers*, slubbings and *roving* to decrease their linear density.

end
A single warp thread through the length of cloth.

epi
Abbreviation for ends per inch.

fell
The edge of the cloth facing the *reed* during weaving where the last *pick* has been put across the warp and beaten up.

ginning
The process which removes cotton fibres from the seed.

handle
The 'feel' of a cloth.

heald (or heddle)
A cotton cord or twisted wire, with loops at both ends, or flat stainless steel strip each with an eye in the centre through which a warp end is threaded. Several healds or heddles are supported on a frame, harness or *shaft*. The frames are manipulated up and down during weaving to form the structure of the cloth.

lags
Wooden slats into which metal or wooden pegs are secured and linked together to form a chain. The chain of lags provides the pattern information to the dobby or witch mechanism which controls the raising of the shafts to form the cloth.

lap
A thick blanket of cotton fibres resulting from the opening and cleaning stages of cotton processing.

lease
A formation at the ends of a warp that maintains an orderly arrangement during warping and preparation processes, and during weaving.

package
A cylindrical object made of wood, metal, paper or plastic, on to which yarn is wound. Bobbins, cones, cops, dobbins, *pirns*, cheeses and spools are all yarn packages.

per cent (%) solution
Refers to the percentage of the active substance contained in a solution.

pick
A single weft thread across the width of the warp.

pirn
A cylindrical object made of wood, paper or plastic on to which weft yarn is wound and then secured in the shuttle for weaving.

pH
A measure of how acid or alkaline a solution is. The result is expressed as a number on a scale of 1-14, with 1 being very acid, 14 being very alkaline, and 7 being neutral. It can be measured with test paper which changes colour according to the pH of the solution.

ppi
Abbreviation for picks per inch.

raschel
A form of warp knitting producing a fabric resembling lace.

reed
Evenly spaced wires, steel strips or bamboo strips held top and bottom between baulks. Used to divide the warp threads in the loom. The reed fits into the *batten, sley* or beater.

reed marks
Lines running through the length of the cloth caused by reed wires when ends are mis-dented.

retting
The process of soaking or fermenting a plant in water or chemicals to loosen the fibres from a woody part of the stem.

roving
The unspun rope of fibres drawn out in the final process before spinning.

scutching
The process of separating the retted fibre from the stem.

selvedges
The edges of woven fabric.

sett
The number of ends and picks in a square unit of cloth.

shaft
A parallel pair of wooden sticks between which are suspended cotton *healds* or a wood or metal frame supporting healds. A set of shafts is known as the harness.

shuttle box
A box at one end of the *sley* from which the shuttle is propelled to a box at the opposite end of the sley during weaving.

size
A starchy substance applied to warp yarns to reduce friction and breakage during weaving.

sley
Also known as a batten or beater, the sley supports the *reed* through which the warp ends are threaded in order. The sley is pushed backwards and forwards by the weaver to beat up the *picks* into the *fell* of the cloth.

sliver
An assemblage of fibres in continuous form without twist.

smash
A breakage in the warp during weaving.

temple
A device used in weaving to ensure that consistent fabric width is maintained.

twaddel (°Tw)
A measure of the relative density of a solution.

union fabric
A fabric made with the warp of one type of fibre and the weft of another.

wale
A column of loops along the length of a knitted fabric

warping creel
A structure for holding yarn packages, usually in tiers and from which an assembly of ends can be withdrawn for warp making.

APPENDIX 7

COMPARISON OF TEX WITH OTHER COUNT SYSTEMS

English cotton count and Tex

COTTON	TEX	COTTON	TEX
1	590	32	18.4
2	296	34	17.4
3	196	36	16.4
4	148	38	15.6
5	118	40	14.8
6	98.4	44	13.4
8	73.8	46	12.8
9	65.6	48	12.4
10	59	50	11.8
11	53.6	52	11.4
12	49.2	56	10.6
14	42.2	60	9.8
16	37	64	9.2
18	32.8	68	8.6
20	29.6	72	8.2
22	26.8	76	7.7
24	24.6	80	7.4
26	22.8	86	6.8
28	21	92	6.4
30	19.6	100	5.9

Woollen counts (Yorkshire skeins woollen) and Tex

WOOLLEN	TEX	WOOLLEN	TEX
6	320	27	72
7	280	28	70
8	240	29	66
10	195	31	62
11	175	32	60
12	160	33	58
13	150	34	56
14	140	35	56
15	130	36	54
16	120	37	52
17	115	38	51
18	107	39	50
19	102	40	48
20	96	42	46
21	92	44	44
22	88	46	42
23	84	48	40
24	80	50	39
25	78	52	37
26	74	56	35

Metric counts and Tex

METRIC	TEX	METRIC	TEX
2	500	60	·16.7
4	250	64	15.6
6	167	68	14.7
8	125	72	13.9
10	100	76	13.1
12	83.3	80	12.5
14	71.4	90	11.1
16	62.5	100	10
18	55.6	110	9.1
20	50	120	8.3
24	41.7	130	7.7
28	35.7	140	7.1
32	31.3	150	6.7
36	27.8	160	6.3
40	25	170	5.9
44	22.7	180	5.6
48	20.8	190	5.3
52	19.2	200	5
56	17.9	210	4.8

INDIRECT FIXED-WEIGHT SYSTEM

English cotton count =
Number of 840-yard hanks per pound

Galashiels woollen count =
Number of 300-yard hanks (cuts) per 24oz

Yorkshire skeins woollen (Y.S.W.) count =
Number of 256-yard hanks per pound

Worsted count =
Number of 560-yard hanks per pound

Linen count =
Number of 1 000-metre hanks per kilogram

DIRECT FIXED-LENGTH SYSTEM

Tex count =
Number of grams per 1 000 metres

Jute count =
Number of pounds per 14 400 yards

Denier count =
Number of grams per 9 000 metres

www.ingramcontent.com/pod-product-compliance
Lightning Source LLC
Jackson TN
JSHW052141131224
75386JS00040B/1311

9781853392115